Writing in Middle School Science

Claim, Evidence, Reasoning Papers that Work

Scott Phillips
The Primal Teacher

CONTENTS

ABOUT THE AUTHOR

Scott Phillips is a veteran teacher with more than ten years of experience teaching middle school math and science. His primal teaching approach is based on presenting information in the best way for student memory retention.

His results in the classroom have produced outstanding state benchmark test scores that are significantly higher than those of his peers. Additionally, his classroom management processes make the teaching job dramatically simpler and easier on the educator.

Scott has a popular blog where he posts free information weekly at www.primalteacher.com. He can also be found on YouTube and Facebook at The Primal Teacher.

INTRODUCTION

Many middle school science classes have been tasked to do Claim, Evidence, Reason writings, which sometimes are called Conclusions Based on Data. Whatever you call them, virtually everyone thinks they are a pain in the neck and most teachers are getting poor to terrible results from the kids.

I used to think they were hard to grade, hard to produce, and of questionable value. By the time you factor in all the time and headaches, I thought they simply were not worth the effort.

At least that's how I used to feel about them. However, now I use them 60-90 times a year in my class and get tremendous results. I can't imagine not using them that often because I have found a way to teach them that makes them easy and amazingly productive as a way to measure student understanding.

In this report, I will show you exactly how I teach CER/CBDs, and I'll include some sample student work for you to see the results I'm getting. Then you can decide for yourself if it's worth trying or not. I think you'll like it once you give it a try.

Please understand that my approach to teaching CERs is going to be different than what you've probably been told to do. And if you are following something close to the methods I was given by three different districts, then you know it's not working very well. You'll need to keep an open mind and carefully review the student work I share at the end of the report. If you can read it in the context of who is doing the work, then I think you'll be just as blown away as I was when the kids starting "getting it."

WHY IT'S NOT WORKING NOW

First, let's start with a couple of the problems you're probably having in your class right now. Do you have the student who just keeps writing and writing and writing and writing a CER in the hopes that if they write enough sentences they will get a good grade? Does your head drop when you see their paper and realize you have to read that hot mess? I know it happened to me many times.

On the flip side, I've had the kid who can't seem even to get started. They turn in either a completely blank page or some nonsensical words strung together with punctuation at the end. Many of my LEP and SPED students produced this kind of paper. However, many students struggle with getting started, and I didn't want this to take all class period, or worse, several periods. I solved both of those problems immediately with the structural approach of my CER. Basically, I teach CERs like I'm a writing teacher instead of a science teacher. Which makes sense because writing teachers are much better at getting good writing samples than are we.

I give my students a very specific structure to follow and several key words to use *every time* they write a CER. I insist the kids write exactly five or six sentences—no more, no less. And I tell them what each sentence should include.

GETTING STARTED

When I first introduce CER writing to my students, I take an entire class period. I walk through an example very slowly and deliberately and explain what each part of the report is designed to do. Here's how I introduce the topic:

"A CER is a way to write like a scientist. It is very structured but it will also make you sound so smart you will really impress your next science teacher. You're going to be so good they'll probably give you an A for the entire class. I just want you to know that this is really important for high school, so we need to learn and practice it now.

"I know some of you have tried CERs before and probably hated them, but you haven't tried it my way. My way is quick and easy. And you'll always know if you got it right because I am going to tell you exactly what to write every time!

"Okay, I can't give you the correct answer, but I can tell you exactly how to structure your report so you get a good grade every time.

"Basically, you're only going to write 5 or 6 sentences—no more, no less. And each sentence has a very specific purpose. Writing is easy when you know your purpose."

Sentence 1:
The Claim

The first sentence in the CER is the claim. I write a chart on the board that outlines the sentences and their purposes, and I fill it in as I explain each one.

SENTENCE	PURPOSE
1. Claim	Answer the question.

I tell them that sentence 1 is the Claim. The purpose of it is to answer the question in one sentence.

The first time you do this with your students, you'll want to work through an example, so they have something concrete to follow. Here's what I use:

Object	Mass (g)	Volume (ml)
Ball	5	10
Rock	1,200	1,000
Elephant	15,555	14,555

And here's what I say:

"Let's say you've been given a data set for several objects.

"The question asks, "Which will float in purified water?"

"You write: "The ball will float." Nothing more, nothing less.

"You don't give any reasons why the ball will float. You don't explain anything at all! Just write, 'The ball will float.' Or, whatever you think will float from looking at the data.

"Don't say 'I think,' 'I believe,' 'probably,' 'maybe,' ever! Don't explain anything. Try to be direct and use the fewest words you can to answer the question.

Sentences 2, 3, 4: The Evidence

The next three sentences provide the evidence for the claim. If the student writes, "The ball will float," then the next sentences describe why that happens. I add these sentences to my chart, and here's what I say:

SENTENCE	PURPOSE
1. Claim	Answer the question.
2. Evidence 1	Support the answer.
3. Evidence 2	Support the answer.
4. Evidence 3	Support the answer.

"In this part of the CER, you are going to provide the evidence for your answer in sentence format. There is some reason you think the ball will float because you read something in the data table or from the description of the experiment. All you do here is change the numbers into sentence format. Pick 2 or 3 data points that support your answer, and write them just like you read them out loud. That's it. You've now already finished the first four sentences. You're two-thirds done!

"In our example you would write, 'The mass of the ball is 5 grams. The volume of the ball is 10 ml. The density of the ball is 0.5 g/ml.'

"That's three sentences right there. You don't need anything else.

"And don't write about any data that doesn't support your claim. If the ball floats, don't write about the rock or elephant, which don't float. They don't help your claim, so leave them out completely."

Sentences 5, 6: The Reasoning

The last two sentences provide the scientific reasoning for the claim. The first sentence states the scientific principle that proves the claim. The second sentence is a simple conclusion statement. Here's how I explain it:

SENTENCE	PURPOSE
1. Claim	Answer the question.
2. Evidence 1	Support the answer.
3. Evidence 2	Support the answer.
4. Evidence 3	Support the answer.
5. Scientific Principle	Proves you are right.
6. Conclusion	Summarizes the evidence and restates the claim

"There are two parts to the reasoning section, and that means exactly two sentences. Sentence five is where you write the scientific principle that shows your claim is right. And we always start it the same way. Write: 'In science we know…'

"Then, state a scientific principle supporting your answer. DO NOT make any reference to the data from the question. The principle must be a generic truth in science and not something specific to the data.

"This is usually the most difficult part of the CER because you may want to write about data from the table. Don't do that. Here's what I mean:

" 'In science we know that things that are less dense than water float in water.' Is that a true statement? Yes. Did I talk about the ball? No. I just used something we know is true in science. Does it prove I am right ABOUT the ball? Yes. But I didn't talk about the ball in this sentence.

"In the last sentence, there are two parts or clauses. They are really easy because all you are doing in this sentence is *rewriting* things you have said already. Then you join the two clauses together with the key word, 'THEREFORE.'

"Start by summarizing your evidence from sentences 2, 3, and 4. With the ball example I would write, 'The density of the ball is less than one.'

"Does that summarize my data? I think it does.

"After that clause, I write the word, 'Therefore.'

"Then, in the second clause, I summarize my claim. For our example I would write, 'It floats.'

"Connecting them all together, I end up with the sentence: 'The density of the ball is less than one, therefore it floats.'

"That's it! You've written a CER!"

Putting it all Together

When I put all of the sentences together in a single paragraph, I have a very concise CER. The question asks, which will float in purified water? My CER says:

"The ball will float in water. The mass of the ball is 5 grams. The volume of the ball is 10 ml. The density of the ball is 0.5 g/ml. In science we know that things that are less dense than water float in water. The density of the ball is less than one, therefore it floats."

How amazing is that? And it's so painless for both students and teachers. Just six easy sentences and students produce a coherent explanation based on given data that is easy to read and grade. I know instantly which of my students understands the basic underlying scientific principle and which may need more help. Using this approach, ALL of my students can now write awesome CERs in less than five minutes and make an A every time!

PRACTICE MAKES PERFECT

When first introducing how to write a CER, I use an entire class period. I walk through writing a CER, as I did in the example above, and then we do another one together as a class. I then have them work two more with a partner, and we go over them in class. Finally, they write two by themselves. That's the entire lesson, and it really works well.

The next day I start class with a CER warm up, and we're off to the races. I use them every time we cover a new scientific principle. I also put up anchor posters around the room to help remind them of the old principles we have covered.

Some of my struggling learners have difficulty with the first couple of CERs, but it doesn't take very long before they are using them really well. Occasionally, the quality of the work will decline for all students. When I see that, I do a quick review of the purpose of each sentence, and student work immediately improves.

SIMPLE GRADING

If you teach students to limit what they write, it forces them to very clearly explain their thinking. I use CERs for warm ups at the beginning of class. I use them on tests and quizzes, too. My students can, after becoming proficient CER writers, complete this assignment in about three minutes. I usually give them four minutes at the beginning while I take roll and get set for class.

I really like CERs because it very quickly shows me student understanding of a concept. A CER also requires that students practice writing and reading—something my LEP and SPED kids need in every subject.

I don't use the complicated three-point grading system that is frequently taught to teachers when learning how to do CERs. I find that to be way too difficult. I usually just read the last sentence to grade the assignment. If sentence six is right and the principle used is correct, the student earns full credit. Otherwise, I start taking off points based on completion.

With my approach, I know if they got it right and if they can connect the science to their answer. That's powerful stuff.

USING CERs

I use CERs all year on every topic. As I said earlier, I have created anchor posters and notes for students to use to identify the proper scientific principles. I identify the four to five most important ideas or principles from any major section (i.e., chemistry, physics, etc.) and then make sure I do several CERs about each of those topics.

In the appendices, I've included several CER prompts along with sample student answers. I've also included the document I give to students when I am teaching CERs. Please feel free to copy it and use it as you wish.

By the way, when I give these assignments, I often have no one miss any of the questions. For the most part, they all use the correct logic for arriving at the answer in spite of the fact that I sometimes try to trick them with a floating car and a sinking balloon. You will also notice the similarity in students' answers. This makes for quick and easy grading and shows me instantly how well they comprehend scientific principles.

And in case you're thinking that this approach is a bit too simplistic, I assure you it's not. I had a high school teacher call me one year to find out what I was doing to get students to write this way. He said my students were noticeably further ahead in their abilities than other students in his class from various schools across the district. That was very nice to hear!

APPENDIX 1:
SAMPLE CERS

Example 1:

Assignment: Write a CER to explain which will float in water?

Object	Density
Water	1.00
Balloon	1.30
Car	0.95
Rock	2.30
Feather	1.10

Student Work:

Here are several examples of CER papers produced by my students. I provided a description of the type of student who did the work to help you evaluate the results.

The car will float.
The density of water is 1.
Every other object is over 1.
ISWK that the less dense object floats.
The car is .95

∴ the car
   will
   float.

This sample is from an on-level 8th grade student. Note that ISWK means "in science we know." Also, note that students use the logic three-dot notation for "therefore."

The car will float.
The Car will float because it ~~is~~ weighs .95 and it is less dense.
In science we know that more dense sinks and less ~~dense~~ floats. less dense than water and
Therefore the car will float in water.

This sample is from an 8th grade student who receives special education services and is also limited English proficient. The student used weight instead of density, but the basic principles are sound.

The car will float   10/38/17
Water is 1.0   B.I.T.W.
Balloon is 1.3
Feather is 1.1
In Science we Know Less dense material will Float
on more dense material
The car is .95

                    The car will Float

This sample is from an 8th grade student who is a level 2 limited English proficient student. This student has great difficulty producing written work in any class.

The car will float in the water.
Water's density is 1. The
car's density is .95. ISWK, less
dense material floats in more
dense material. The car's density
is .95, therefore, it floats.

This sample is from a 7th grade student who is autistic and dyslexic.

Example 2:

Assignment: Write a CER to answer the following question: Which reactions demonstrate the Law of Conservation of Mass?

	Reactants		Products
Decomposition of Hydrogen Peroxide	$2H_2O_2$	→	$2H_2O + O_2$
TOTAL Molar Mass	68.1292g		68.0292g
Composition (synthesis) of sodium chloride (table salt)	$2Na + Cl_2$	→	$2NaCl$
TOTAL Molar Mass	116.8854g		116.8854g
Single replacement reaction	$Mg + 2HCl$	→	$MgCl_2 + H_2$
TOTAL Molar Mass	98.2268g		97.2268g

I like this kind of CER because it forces the kids to read very carefully. They also have to recognize that the equations look balanced according to their formulas but are not when you look deeper to the mass. Getting to that second level of thinking is very important and difficult to measure unless you use something like a CER.

Student Work:

1. Total Molar mass is demonstarting the
L.O.C.M..
2. The reactant is 116.8854g.
3. The product is 116.8854g.
4. The others are wrong.
5. In Science we know that L.O.C.M. says
stuff goin will come out.
6. Total Molar mass is demonstrating the L.O.C.M
therefore, it is correct.

This sample is from a 7th grade on-level student.

$2Na + Cl_2 \rightarrow 2NaCl$ demonstates
the law of conservation of Mass.
The reactans are 2 Na and 2 Cl.
The at the beging mass 116.8854g.
mass
At the end the mass was 116.8854g.
In Science we know the law of conservation
of mass says what goes in must come
out. The mass stays the same. therefore it follows
the law of conservation of mass.

This sample is from a 7th grade on-level student.

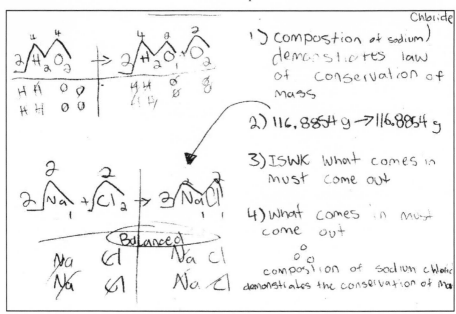

This sample is from a 7th grade student who has limited English proficiency and behavior issues. He has a great deal of difficulty starting work in all classes.

Example 3:

Assignment: Magic Coins

A student saw the movie "The Alchemist" and wanted to duplicate the experiments he saw on the screen. He was particularly interested in making silver coins to sell to collectors. In the movie, the star could mix frog legs, eye of newt, a pinch of wolf's bane, some salt peter, and the tears of 200 angry goats to produce a potion. This potion was placed in a cauldron and stewed with 11 pine cones, 5 mandrake roots and 17 black widow spiders for a fortnight until it was good and frothy. Then, under the silver moonlight of a full moon, the potion was placed in molds shaped like coins. Once covered by the branches of a birch tree and stimulated by the screech of a horned owl, the silver coins would be produced.

Please write a CER answering the following question: Can the boy make silver coins using this method?

(NOTE: This was a CER I did early in the year. I gave the students a template to use to help prompt their thinking.)

Student Work:

Claim: The method wouldn't work to make silver coins.

Evidence: It has no silver. Yet, it has molds for the coins.

Reasoning: (Hint: Law of Conservation of Mass)

In Sciene we know that what goes in must come out. There is no silver inside therefore no silver coins are made.

This sample is from a 7th grade pre-AP student.

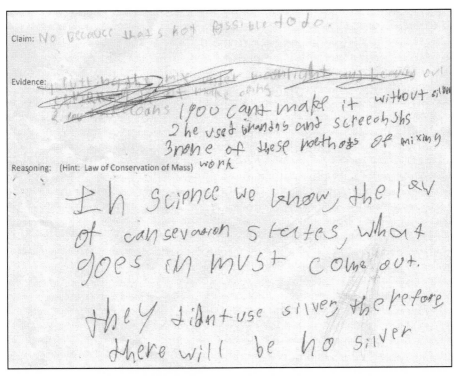

Claim: No because that's not possible to do.

Evidence:
1 you cant make it without silver
2 he used brandnb and screech shs
3 none of these methods of mixing work

Reasoning: (Hint: Law of Conservation of Mass)

In science we know, the law of conservation states, what does in must come out. they didnt use silver, therefore there will be no silver

This sample is from a 7th grade student who receives special education services and has limited English proficiency. This student has nearly on-level verbal skills but has great difficulty with written expression.

Claim: I don't think he will be able to make a silver coin.

Evidence: Silver Coins have to be made with silver.

Reasoning: (Hint: Law of Conservation of Mass)

Silver coins need silver and its not listed in the ingredients.

In science we know you need an element for an element to be made. Therefore the boy can not make silver coins using this method.

This sample is from a 7th grade student who has limited English proficiency. Note the use of "I don't think" in the claim. However, as stated previously, this was done early in the year. The science is sound.

APPENDIX 2:
CER SIX-SENTENCE
STRUCTURE HANDOUT

Claim/Evidence/Reasoning
Six-Sentence Structure

This is the information I provide to students as a handout.

Sentence 1: Answer the question

Sentences 2, 3, 4: Convert data from number format into sentence format. Make sure to exclude unimportant data.

Sentence 5: Start with "In science we know…" then state the scientific principle that supports your answer.

Sentence 6: Summarize data then write "therefore" and summarize your answer.

You should never write more than six sentences on my CERs. You should never use the word "because."

Example:

Question: Will the ball sink in water?

Object	Density
Ball	1.5 g/ml
Water	1.0 g/ml
Corn	0.25 g/ml

CER:
1. Yes, the ball will sink.
2. The density of the ball is 1.5 g/ml.
3. The density of water is 1.0 g/ml.
4. In science we know that a more dense object will sink in a less dense material.
5. The ball is denser than water, therefore it sinks.

Notice I didn't include any information about corn because it doesn't support the answer.

You could write the CER in paragraph form like this:

Yes, the ball will sink. The density of the ball is 1.5 g/ml. The density of the water is 1.0 g/ml. In science we know that a more dense object will sink in a less dense material. The ball is denser than water, therefore it sinks.

In this example, we only used five sentences, but the idea is the same. You would use more data to support your answer if it was necessary.

Here's a question like that.

Example:

Which will float on top?

Object	Density
Ball	0.5 g/ml
Water	1.0 g/ml
Oil	0.75 g/ml

Here, you need the data from all three objects to answer the question.

ABOUT PRIMAL TEACHING

My name is Scott Phillips, and I am a primal teacher. We are a growing community of educators who believe the American education system is as broken as the American diet.

If you're a teacher, you know the problems we face. Out of control students, dismal test scores, and outrageous teacher turnover to name just a few. But this is not a movement to overhaul the American educational system if that is even possible. Instead, we seek to make an impact where it counts – in our classrooms.

Primal teachers seek to fix our schools by making simple changes in the classroom. This organic and natural approach to learning improves student performance and streamlines the teaching process. Our goal is to help teachers succeed and get the results they are tirelessly working to achieve.

I wrote *The Primal Teacher: A Caveman's Secrets to Improving Your Class* to share ideas that worked for my students and me. I present strategies in the book culminating from more than ten years of teaching middle school math and science to struggling learners. During that time, I tried new activities, challenged old ideas, and kept tweaking and fine-tuning until I got the results I wanted.

My student performance on state benchmark tests is significantly and consistently higher than those of my peers. Additionally, these strategies make the teaching job dramatically more straightforward and easier on the educator.

So, why do I call it Primal Teaching? Because it's all about getting back to basics and using results-based ideas that work.

The primal teaching movement recognizes that people have been learning the same way for a very, very long time. We know there are some biological truths about the way in which our brains learn new information, and primal teachers don't try to fight nature.

A primal teacher believes in the importance of teaching through pattern recognition and the power of scheduled reviews. We also recognize that humans spent tens of thousands of years educating themselves through storytelling. It's a natural part of who we are, and primal teachers believe storytelling is perhaps the most effective and overlooked teaching modality ever invented.

If you are an educator in any capacity, I hope you will join us. You can follow my blog at primalteacher.com where I offer ideas for simple and easy ways to deal with classroom issues. And I want to hear your thoughts in return! You can also subscribe to my YouTube channel and follow me on Facebook. Just search for The Primal Teacher.

Thanks, and remember to make every day epic!

Scott Phillips, Caveman

P.S. I hope you will consider doing a review for this book on Amazon. It would really support my efforts. Thank you!

Made in the USA
Las Vegas, NV
05 April 2023

70233290R00023